Forest Conservation

Tom walked in to see Emma, Eva and George dancing around with animal masks on. `They are playing Jungle Book yet again,´ laughed Beth, their mother.

`Then I have a surprise for all! Who wants to be my jungle trooper?´ said Tom, waving the tickets to the Forest Reserve.

'Me! Me! Me!' shouted all three together. 'Well, early to bed then, if you don't want to be left behind,' smiled Beth.

At dawn, the excited jungle troopers drove away from the city, up towards the hills.

Suddenly, Tom turned a bend and stopped in front of a waterfall.

`What a lovely picnic spot! Look at all the trees and butterflies! And hear the chirping of birds!' Eva clapped her hands. Having dipped their feet in the cold, clear water, they munched on sandwiches, looking at the dense forest around.

'Children, did you know that forests cover more than 30 percent of the land on our planet?' Beth asked the children.

'Wow, why do we need so much of forest, Mom?' asked Emma.

'Forests are home to many animals, insects and birds. Sometimes, even people. But forests have many other uses as well!' Beth replied.

'Forests have trees which help us to breathe. Can you guess how?' Beth asked the children.

'My teacher said that trees produce the oxygen that we breathe in and absorb the harmful gases that we breathe out,' Emma showed off her knowledge.

'Correct, Emma! And they also take care of our climate by keeping the Earth cool. Forests help in making rain. If it rained and there were no forests, we would have floods and all the good soil would be gone,' Beth told them.

'Alright, break is over... Forest Camp, here we come,' jumped up Tom.

'Look, that squirrel just picked some nuts from that tree,' yelled out George. 'And I spotted some mushrooms, Daddy,' Eva exclaimed.

'Forests give us food like nuts, berries, fruits, seeds, and mushrooms. But remember, not all of them can be eaten by us.'

'Forests also give us wood for our homes, furniture and fuel,' added Emma.

Beth smiled, 'Yes, paper is made from wood pulp. Fibre from cotton, jute, and even bamboo, is used to make cloth. Also, different parts of trees and plants like the roots, bark, leaves are used to make medicines.'

Sighting the forest camp ahead, Tom looked at the excited faces and said, `Now some ground rules. Don't play with fire or wander off on your own. We'll go hiking after some food and rest.´

'Look Dad, what beautiful deer! But why are they eating plastic?' asked Eva, as Tom led them down a walking trail.

'This happens when people litter forests with plastic and trash. It can kill these animals. If you see any litter around, put it in this trash bag,' Tom suggested.

'Everybody, come here,' Emma cried aloud from the corner ahead. 'Where are all the trees? There are only stumps here!'

`People often burn large areas of forest to construct houses, build hotels, roads, etc., without thinking of replanting the trees. People also cut down trees for the goods we use,´ explained Beth sadly.

`We cut 26 trees to make just one ton of paper. And globally we use hundreds of million tons of paper each year. There is no end to our greed. But if we keep taking from the forest without giving back to it, it is we who will suffer the most,´ she added.

'At this rate, we will soon lose all our forests, birds and animals,' cried George angrily. 'Why are such people not punished?'

'The answer to this is Forest Conservation or protection of forests. We do have strict laws to punish those who harm the forests and the wildlife in it,' Beth comforted him.

'Conservationists are the people who work hard to save forests. They stop the unplanned cutting of forests and plant more trees. They prevent accidental forest fires and create walking trails to prevent visitors from harming the forests,' added Tom.

'What is this?' curious Emma pointed to a pit in the camp. 'The sign says 'Rain Harvester'.'

Tom smiled, 'Protecting our forests also means conserving water bodies. In the forests, trees and plants break the fall of rain, and water seeps gently into the ground, slowly feeding our wells, lakes and rivers. But with climate change and less rainfall, especially in the hills, we need to collect or harvest rainwater, use water wisely, and even reuse it to avoid wastage.'

'But Mom, how can we help conserve forests?' asked a worried Eva.

'Well, Eva, we can start by using less paper or use paper wisely like recycling and using both sides of scrap paper for your drawings. Also share books amongst friends and donate the ones you no longer need,' Mom replied.

'That's a super idea!' George exclaimed.

'Planting trees regularly is another way. Did you know that at the camp you can plant trees in your name as a parting gift to this wonderful forest?'

'Yay!' shouted all three. 'And years later, we can come and see our own grown up trees!'

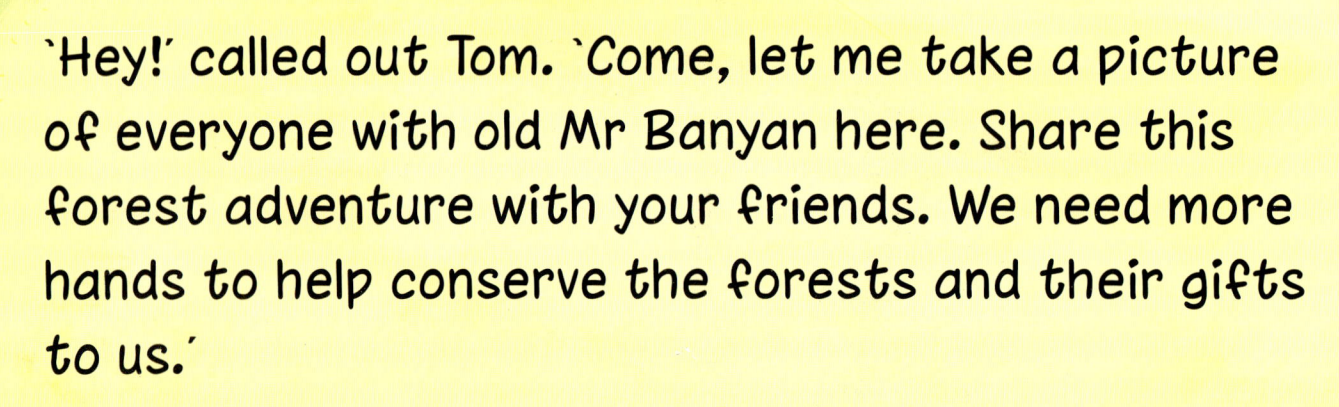

'Hey!' called out Tom. 'Come, let me take a picture of everyone with old Mr Banyan here. Share this forest adventure with your friends. We need more hands to help conserve the forests and their gifts to us.'

'Yes Dad. Save Forests, Save Earth! We are the Forest Patrollers!' all three said in chorus.